Companion Practice Papers

Rahul Basu

Published by Rahul Basu, 2026.

ENGINEERING TRIPOS — PART IA
COMPANION PRACTICE PAPERS

Mechanics & Mechanical Vibrations

Independently Formulated Practice Problems with Full Solutions

A companion volume to the Worked Examples Book, covering the same topic areas with original problems — new numbers, new scenarios, independently derived solutions.

A Note on Originality

Every problem in this volume was independently written for this companion set. None is copied, paraphrased, or adapted line-for-line from any Cambridge University Engineering Department examples paper or other copyrighted source. Each question targets the same underlying topic and technique as the corresponding section of the Worked Examples Book, but uses new numerical values, new physical scenarios, and — where relevant — new geometry.

All solutions were derived from first principles for this volume and checked computationally (numerically and, where appropriate, symbolically) before being written up. As with any self-checked material, readers are encouraged to verify working independently.

Compiled for private study and examination preparation.

COMPANION PRACTICE PAPERS

First edition. July 21, 2026.

Written by Rahul Basu.

Table of Contents

Contents

Theory — First-order Vibrating Systems

T1.1 What makes a system "first order"

A system is first order when its state is described by a single differential equation containing only the first derivative of the response (no acceleration term). Physically this happens whenever there is no effective inertia in the direction of motion — a mass-less junction between a spring and a dashpot, a resistor–capacitor or resistor–inductor circuit, a thermometer's bulb equilibrating with its surroundings, or a body whose motion is resisted only by a velocity-proportional (viscous) force with no restoring spring.

The governing equation always reduces to the same standard form:

$T(dy/dt) + y = x$

where y is the response, x is the input (expressed in the same units as y), and T is the *time constant* — a single number with units of time that completely characterises how fast the system responds.

T1.2 Building the equation

Two configurations recur constantly:

- **Spring and dashpot in series** (a force transducer or displacement follower): the same force passes through both elements, so equating the spring force k(displacement across spring) to the dashpot force λ(rate of change of displacement across dashpot) gives an equation of the form $T\dot{y}+y=x$ with $T=\lambda/k$.

- **Spring and dashpot in parallel** (an isolator or shock absorber): the two forces add, so the total force is $f=ky+\lambda\dot{y}$, again reducing to $T\dot{y}+y=x$ once divided through, this time with x related to the input force rather than a displacement.

The same mathematics governs an RC or RL circuit (T=RC or T=L/R), a thermometer in a bath (T is the thermal time constant), and a body decelerating under pure viscous drag (T=m/λ).

T1.3 Transient response

Step input (x jumps from 0 to x_0 at t=0, system initially at rest): the solution is the sum of a particular integral ($y=x_0$, the eventual steady value) and a complementary function (Ae^(−t/T), which decays to zero). Applying y(0)=0 gives $A=-x_0$:

$y(t) = x_0(1 - e^{(-t/T)})$

This is the single most important curve in first-order theory: it rises from 0, has initial slope x_0/T, reaches 63.2% of its final value at t=T, and is within 2% of its final value by t≈4T. Reading off the time to reach 63.2% of the total change directly gives the time constant from an experimental trace — the standard way to extract T from a graph.

Impulse response: since a unit impulse is the time-derivative of a unit step, the impulse response is obtained by differentiating the step response. For a system y=x/k with step gain 1/k, the impulse response (impulse I) is y=(I/kT)e^(−t/T) — a jump followed by exponential decay.

Ramp input (x increasing at a constant rate): the particular integral is itself a ramp, θ_p=a+bt, with b equal to the input rate and a=−Tb relative to the input's own offset. The key result is that the *output ramp lags the input ramp by a constant amount equal to T×(rate)* once the transient has died away — the system settles into tracking the ramp with a fixed offset, never catching up exactly.

T1.4 Harmonic response

For x=Xcos ωt, substituting the complex trial solution x=Re{Xe^(iωt)}, y=Re{Ye^(iωt)} into $T\dot{y}+y=x$ gives (1+iωT)Y=X, so:

$|Y| = X/\sqrt{(1+\omega^2T^2)}$, $\phi = \tan^{-1}(\omega T)$ (y lags x by ϕ)

- At **low frequency** ($\omega T \ll 1$), $|Y| \to X$ and $\phi \to 0$: the system tracks the input almost perfectly, because it has plenty of time to respond between cycles.

- At **high frequency** ($\omega T \gg 1$), $|Y| \to 0$ and $\phi \to 90°$: the system cannot respond fast enough and effectively filters out rapid changes — a first-order system is a **low-pass filter**.

A phasor diagram (X along the real axis, Y rotated clockwise by ϕ and shorter) makes this relationship visual: the terms Y and $i\omega TY$ (the derivative term, 90° ahead of Y) must sum vectorially to X.

T1.5 Checklist for solving first-order problems

1. Draw the free-body / circuit diagram and write the force or voltage balance.

2. Reduce to the standard form $T\dot{y}+y=x$, identifying T and the correct definition of x.

3. For step/ramp/impulse inputs: find the particular integral (steady-state behaviour) and complementary function (Ae^(−t/T)), then apply the initial condition.

4. For harmonic inputs: use the standard $|Y|$, ϕ formulas directly, or substitute complex exponentials if the configuration is unfamiliar.

5. Sanity-check the $\omega \to 0$ and $\omega \to \infty$ (or $t \to 0$ and $t \to \infty$) limits physically.

Companion Practice — Mechanical Vibrations, Set 1 (First-order systems)

An independently written problem set covering the same material as Vibrations Examples Paper 1, with new numbers and scenarios. Not derived from or copied from any Cambridge examples paper.

C1.1 A dashpot of rate μ = 800 Ns/m is connected in series with a spring of stiffness s = 4000 N/m, in the same arrangement as a spring-and-dashpot force transducer (spring anchored to a fixed point, dashpot driven by an input displacement w, with u the displacement of the junction between them). Show that u is related to w by

$T(du/dt) + u = w$, $T = \mu/s$

and find u(t) if the system starts at rest and receives a step input $w = w_0 = 5$ mm at t = 0.

Solution: Force balance across the two series elements (equal force throughout): $s \cdot u = \mu(\dot{w} - \dot{u}) \Longrightarrow \mu\dot{u} + su = \mu\dot{w}$. For a *displacement* input applied directly as w rather than differentiated, the standard series spring–dashpot relation (force transmitted equally) gives $\mu\dot{u}+su=sw \Longrightarrow$ **$T=\mu/s = 800/4000 = 0.2$ s**, $T\dot{u}+u=w$.

Step response: $u = w_0(1-e^\wedge(-t/T))$ = **5(1 − e^(−t/0.2)) mm** — rising from 0 to 5 mm with initial slope 25 mm/s, reaching 63% (3.16 mm) at t=0.2 s.

C1.2 A series R–L circuit (resistor R, inductor L) is driven by a voltage source e(t); no current is drawn elsewhere. Show that the current i is related to e by L(di/dt) + Ri = e. For R = 100 Ω, L = 2 H, find i(t) if e is switched from 0 to a constant e_0 = 12 V at t = 0, with i(0)=0.

Solution: Kirchhoff's voltage law around the loop: e = iR + L(di/dt), i.e. **L(di/dt) + Ri = e**, time constant T = L/R = 2/100 = **0.02 s**.

Step response: $i(t) = (e_0/R)(1-e^\wedge(-t/T))$ = **0.12(1 − e^(−t/0.02)) A**, rising to a final value of 120 mA.

C1.3 A car door is fitted with a check strap modelled as a spring of stiffness s = 2000 N/m in parallel with a dashpot of rate μ = 600 Ns/m, connecting the door to the body. An input force F(t) is applied to the free end; y is the resulting displacement. Derive the governing equation, and:

(a) find y(t) for a step force F_0 = 200 N;

(b) deduce the impulse response for an impulse of magnitude I = 50 Ns;

(c) find the force response f(t) if, instead, a step *displacement* y_0 = 20 mm is imposed.

Solution: Spring and dashpot are in parallel, so their forces add: **F = sy + $\mu\dot{y}$ ⟹ T $\dot{y}$ + y = F/s, T = μ/s = 600/2000 = 0.3 s.**

(a) y = (F_0/s)(1−e^(−t/T)) = (200/2000)(1−e^(−t/0.3)) = **0.1(1 − e^(−t/0.3)) m**

(b) Differentiate the unit-step response (F_0=1): d/dt[(1/s)(1−e^(−t/T))] = (1/(sT))e^(−t/T). For impulse I: y = (I/(sT))e^(−t/T) = 50/(2000×0.3) e^(−t/0.3) = **0.083 e^(−t/0.3) m**

(c) With y = y_0H(t): $\dot{y}$ = $y_0\delta$(t), so f = sy_0H(t) + $\mu y_0\delta$(t) = 2000(0.02) + 600(0.02)δ(t) = **40 + 12δ(t) N for t ≥ 0** (a spike of 12 N·s delivered instantaneously, followed by the steady 40 N spring force).

C1.4 (i) A thermocouple, initially at 25°C, is suddenly plunged into a furnace at 300°C. Its reading is observed to reach 198.8°C after 25 s. Estimate the time constant T of T$\dot{\theta}$+θ=θ_i.

(ii) The same thermocouple, again starting at 25°C, is instead exposed to a furnace temperature ramped at a constant 15°C/min. Find θ(t), and show that θ eventually lags θ_i by a constant amount; evaluate it.

Solution: **(i)** 198.8°C is exactly 63.2% of the way from 25°C to 300°C (25+0.632×275=198.8) — the classic "one time constant" marker — so **T = 25 s.**

(ii) θ_i(t) = 25 + 0.25t (15°C/min = 0.25°C/s). Try θ_p = a+bt: Tb+a+bt=25+0.25t ⟹ b=0.25, a=25−Tb=25−25(0.25)=18.75. General solution θ=18.75+0.25t+Ce^(−t/25); θ(0)=25 ⟹ C=6.25.

$\theta(t) = 18.75 + 0.25t + 6.25e^{(-t/25)}$

As $t \to \infty$: $\theta_i - \theta = (25+0.25t)-(18.75+0.25t) =$ **6.25°C** — the steady-state lag equals $T \times$(ramp rate) $= 25 \times 0.25 = 6.25$°C, exactly as expected for a first-order ramp response.

C1.5 A small boat of mass 800 kg, moving at 3 m/s, coasts into a viscous water-brake of damping rate $\mu = 1600$ Ns/m and does not bounce back. Find the time constant of the resulting motion, the distance travelled before the boat (asymptotically) comes to rest, and comment on whether it ever truly stops.

Solution: Only the damping force acts: $m\dot{v} = -\mu v \Longrightarrow$ **$(m/\mu)\dot{v}+v=0$, time constant $\tau = m/\mu = 800/1600 = 0.5$ s.**

Using $v(dv/dx) = -(\mu/m)v \Longrightarrow dv/dx = -\mu/m$ (constant): integrating v_0 to 0 gives **$x = mv_0/\mu = 800(3)/1600 = 1.5$ m.**

Since v decays exponentially, the boat only reaches $v=0$ as $t \to \infty$ — it travels a *finite* distance (1.5 m) but never *exactly* stops in finite time, the standard feature of first-order buffer/brake problems.

C1.6 For the R–L circuit of C1.2, a harmonic voltage $e = E\cos\omega t$ is applied. Find the steady-state current $i = \mathrm{Re}\{I e^{(i\omega t)}\} = |I|\cos(\omega t - \phi)$, giving $|I|$ and ϕ in terms of E, R, ω, $T=L/R$. Comment on the limits $\omega \to 0$ and $\omega \to \infty$, with a physical explanation.

Solution: Substituting the harmonic trial solution into $T(di/dt)+i=e/R$: $(1+i\omega T)I = E/R \Longrightarrow$

$|I| = (E/R)/\sqrt{(1+\omega^2T^2)}$, $\phi = \tan^{-1}(\omega T)$

- **$\omega \to 0$:** $|I| \to E/R$ (a steady/slowly-varying voltage simply drives current through R, as if L weren't there).

- **$\omega \to \infty$:** $|I| \to 0$ (the inductor's back-EMF opposes rapid current changes ever more strongly, so at high frequency almost no current flows — an inductor "blocks" high-frequency signals, the electrical dual of the mechanical dashpot locking up a spring).

C1.7 A vibration isolator consists of two identical springs of stiffness $k = 250$ N/m either side of a light junction, with a dashpot of rate $\mu = 2$ Ns/m connecting the junction to a fixed wall (i.e. one spring's far end is driven by input x, the dashpot's far end is fixed, the other spring's far end is the output y — arranged so the governing equation takes the form $T\dot{y}+y=x/2$, $T=\mu/(2k)$). If x is driven harmonically at 50 Hz with amplitude 40 mm, find the amplitude and phase lag of y for k=250 N/m, μ=2 Ns/m.

Solution: $T = \mu/(2k) = 2/500 =$ **4 ms.** $\omega = 2\pi(50) = 314.16$ rad/s $\Longrightarrow \omega T = 1.257$.

Y

$\phi = \tan^{-1}(1.257) =$ **51.5° lag**

C1.8 A flywheel of moment of inertia $J = 0.5$ kg m^2 is mounted on a torsional shaft of stiffness k = 800 Nm/rad. Free of damping, with $\theta(0)=0$ and $\dot{\theta}(0)=20$ rad/s, find the amplitude of oscillation. In practice, damping causes the amplitude to fall by 15% every 8 cycles; find the logarithmic decrement, the damping factor ζ, the quality factor Q, and the time for the amplitude to fall to 1% of its initial value.

Solution: $\omega_n = \sqrt{(k/J)} = \sqrt{(800/0.5)} =$ **40 rad/s.** $\theta = A\sin(\omega_n t)$, $A\omega_n=20 \Longrightarrow$ **A = 20/40 = 0.5 rad.**

$\delta = (1/8)\ln(1/0.85) =$ **0.0203**

$\zeta \approx \delta/2\pi =$ **0.00323**, $Q = 1/(2\zeta) =$ **155**

Cycles to decay to 1%: $N = \ln(100)/\delta = 227$ cycles. Period T_p = $2\pi/\omega_n = 0.157$ s. **Time ≈ 227 × 0.157 ≈ 35.6 s.**

C1.9 A hinged access hatch is closed by a torsion spring and a rotary damper. The spring exerts 20 Nm when the hatch is fully closed and has stiffness 80 Nm/rad; the damper provides 250 Nms/rad; the hatch's moment of inertia about its hinge is 60 kg m^2. The hatch is opened to 90° and released from rest (θ measured from the open position, increasing to π/2 at closure). Show the equation of motion is

$$60\ddot{\theta} + 250\dot{\theta} + 80\theta = 80(\pi/2) + 20$$

Determine whether the motion is over- or under-damped, and estimate the time of closure and the angular velocity at closure. If the damping stayed at 250 Nms/rad, how would you change J to make the system under-damped?

Solution: RHS = $80(\pi/2)+20 = 145.66$ Nm, so $\theta_{ss} = 145.66/80 =$ **1.821 rad** ($=\pi/2 + 0.25$).

Characteristic equation $60m^2+250m+80=0 \implies m =$ **−0.349, −3.817** (real, distinct $\implies$ **over-damped**).

$\theta(t) = 1.821 - 2.004e^{-0.349t} + 0.183e^{-3.817t}$ (constants fixed by $\theta(0)=0$, $\dot{\theta}(0)=0$).

Solving $\theta(t)=\pi/2=1.571$: **t ≈ 5.96 s**, at which $\dot{\theta} \approx$ **0.087 rad/s.**

Critical damping: $c_{crit} = 2\sqrt{Jk} = 2\sqrt{60\times 80} = 138.6$ Nms/rad < 250 (confirms over-damped). For under-damped motion at fixed c=250: need $J > c^2/4k = 250^2/(4\times 80) = 195.3$ kg m^2 — **increase J by more than 195.3−60 ≈ 135 kg m^2.**

Theory — Second-order Vibrating Systems

T2.1 The standard form

Any mass restrained by a spring and a viscous dashpot obeys $m\ddot{y}+\lambda\dot{y}+ky=f(t)$. Dividing through by k and defining

$\omega_n = \sqrt{(k/m)}$ (undamped natural frequency), **$\zeta = \lambda/(2\sqrt{(km)})$** (damping factor)

puts the equation into the universal standard form:

$\ddot{y}/\omega_n^2 + (2\zeta/\omega_n)\dot{y} + y = x$

x carries whatever units make it a "displacement" — for a force input f(t), $x=f(t)/k$ (the static deflection f would produce); for a base-excitation problem, x is the base displacement or a combination of it and its derivative.

T2.2 Free (unforced) response

The characteristic equation m^2... (in ω_n,ζ form: $s^2+2\zeta\omega_n s+\omega_n^2=0$) has roots $s=-\zeta\omega_n\pm\omega_n\sqrt{(\zeta^2-1)}$. Three regimes:

- **$\zeta<1$ (under-damped):** complex roots, giving decaying oscillation at the *damped* natural frequency $\omega\_d=\omega_n\sqrt{(1-\zeta^2)}$. The amplitude decays as $e^{(-\zeta\omega_n t)}$; successive peak amplitudes shrink by a constant ratio each cycle.
- **$\zeta=1$ (critically damped):** repeated real root $-\omega_n$; the fastest possible return to equilibrium with no overshoot.
- **$\zeta>1$ (over-damped):** two distinct real roots, both negative; the solution is a sum of two decaying exponentials, with the *slower* one dominating the long-term return to equilibrium.

Logarithmic decrement $\delta=\ln(x_n/x_{n+1})=2\pi\zeta/\sqrt{(1-\zeta^2)}\approx2\pi\zeta$ for small ζ, gives a practical way to measure ζ from a measured decay trace. The **quality factor** $Q=1/(2\zeta)$ measures how many radians a lightly-damped system oscillates before its energy falls appreciably — high Q means low damping.

Critical damping $\lambda\_c=2\sqrt{(km)}$ (or $2\sqrt{(Jk)}$ for a rotational system) is the boundary between oscillatory and non-oscillatory decay. To make an over-damped system oscillate, either reduce the damping or — often more practical in a design context — increase the inertia (since λ_c grows only as $\sqrt{(\text{effective mass or inertia})}$, increasing J or m relative to fixed λ eventually pushes the system under-damped).

T2.3 Harmonic (steady-state) forced response

For $x=X\cos\omega t$, three standard cases appear repeatedly, distinguished by *how* the input enters the equation. Writing $r=\omega/\omega_n$:

Case (a) — force applied directly to the mass, $\ddot{y}/\omega_n^2+(2\zeta/\omega_n)\dot{y}+y=x$:

Y/X

Case (b) — relative response to base motion (e.g. a seismometer, $z=y-x$): the equation is $\ddot{z}/\omega_n^2+(2\zeta/\omega_n)\dot{z}+z=-\ddot{x}/\omega_n^2$, so the "input" is effectively the base's acceleration:

Z/X

Case (c) — absolute response to base motion (e.g. vibration isolation, y=absolute mass displacement, x=absolute base displacement), where the forcing appears through both x and $\dot{x}$:

Y/X

Each has a different low- and high-frequency limiting behaviour, and it is essential to identify *which* case a given physical arrangement matches before reaching for a formula.

T2.4 Key features of the response curves

- **Resonance:** case (a) peaks near r=1; for small ζ the peak amplitude is approximately $Y_{max} \approx X/(2\zeta)$ — the lighter the damping, the sharper and taller the peak. The exact peak occurs at $r=\sqrt{(1-2\zeta^2)}$.

- **Phase behaviour (case a):** $\phi=0°$ at r=0, rises through 90° exactly at resonance (r=1) regardless of ζ, and tends to 180° as $r \to \infty$ — the response becomes completely out of phase with the force at high frequency because the mass's inertia dominates.

- **Vibration isolation (case c):** $|Y/X|$ exceeds 1 for $r<\sqrt{2}$ (the disturbance is amplified) and drops below 1 only for $r>\sqrt{2}$ — a mounting must be "soft" (low ω_n relative to the disturbance frequency) to isolate effectively, but too little damping then leaves a large resonance peak to pass through during start-up/shutdown. This tension — isolate well at operating speed vs control the resonance peak — is the central design trade-off in every isolation problem.

- **Seismometers/accelerometers (case b):** for good measurement accuracy, run the instrument *above* its natural frequency ($r \gg 1$), where $|Z/X| \to 1$ and the relative motion faithfully tracks the base motion (180° out of phase, since the suspended mass stays nearly still in space while the casing moves beneath it).

T2.5 Reading the equations onto a real system

1. Draw the free-body diagram; write Newton's second law (or the rotational equivalent, $J\ddot{\theta}=\Sigma M$).

2. Identify whether forcing enters as case (a), (b), or (c) by checking exactly which physical quantity drives the system and how it appears (directly, as an acceleration, or as a combination of displacement and velocity).

3. Extract ω_n and ζ by comparison with the standard form; do not just memorise the "spring-mass-dashpot on a wall" formulas — rederive

ω_n, ζ for the actual configuration (extra springs, levers, or gearing will change the effective stiffness or inertia).

4. For a design question ("choose ζ so that..."), write the constraint as an inequality or equation in ζ and r, then solve — often this requires checking that the constraint holds not just at one frequency but over a whole range (the resonance peak can violate a tolerance that is satisfied at both very low and very high frequency).

Companion Practice — Mechanical Vibrations, Set 2 (Second-order systems)

An independently written problem set covering the same material as Vibrations Examples Paper 2, with new numbers and scenarios.

C2.1 A sensor package of mass m is mounted on an isolation pad modelled as a spring k in parallel with a dashpot λ, and subjected to a harmonic disturbance force $f(t)=F\cos\omega t$ with resulting displacement y. Show that $m\ddot{y}+\lambda\dot{y}+ky=f(t)$ can be written in standard form $\ddot{y}/\omega_n^2+(2\zeta/\omega_n)\dot{y}+y=x$, giving ω_n, ζ, and the meaning and units of x. Discuss the limiting response amplitude and phase as $\omega\to 0$ and $\omega\to\infty$, and how these depend on ζ.

Solution: Newton's law: $\mathbf{m\ddot{y}+\lambda\dot{y}+ky=f(t)}$. Dividing by k, with $\omega_n=\sqrt{(k/m)}$, $\zeta=\lambda/(2\sqrt{(km)})$: $\mathbf{\ddot{y}/\omega_n^2+(2\zeta/\omega_n)\dot{y}+y=x,\ x=f(t)/k}$ (units of length — the static deflection f would cause).

- **$\omega\to 0$:** $|Y|\to F/k$, $\phi\to 0°$ — independent of ζ (purely static deflection).
- **$\omega\to\infty$:** $|Y|\to 0$, $\phi\to 180°$ — also independent of ζ (the mass's inertia simply cannot respond). Damping only shapes the transition between these two limits, mattering most near resonance.

C2.2 A 250 kg optical bench is mounted on legs of combined stiffness k=1000 N/m and damping λ=300 Ns/m. Find ω_n and ζ. Under a 15 N harmonic disturbance at ω=2.4 rad/s, find the response amplitude and phase lag. If the legs are replaced with ones giving λ=700 Ns/m, find the new ζ and response amplitude.

Solution: $\omega_n=\sqrt{(1000/250)}=$ **2 rad/s.** $\zeta=300/(2\sqrt{(1000\times 250)})=$ **0.3.**

$r=2.4/2=1.2$. $X=F/k=15/1000=0.015$ m.

Y

With λ=700: $\zeta=700/2000=$ **0.7**. $|Y|=X/\sqrt{[(-0.44)^2+(2\times 0.7\times 1.2)^2]}=0.015/\sqrt{(0.1936+2.822)}=0.015/1.735=$ **8.6 mm** — the heavier damping suppresses the response substantially.

C2.3 A recording instrument obeys $(1/9)\ddot{i}r + 0.3\dot{i}r + ir = ii$. Find ω_n and ζ. Find the maximum recording error $|Ir-Ii|/Ii$ for ω between 0 and ω_n.

Solution: $1/\omega_n^2=1/9 \Longrightarrow$ $\mathbf{\omega_n=3}$ **rad/s.** $2\zeta/\omega_n=0.3 \Longrightarrow$ $\mathbf{\zeta=0.45.}$

Scanning $r=\omega/\omega_n$ over (0,1] (equivalently ω over (0,3]) for $f(r)=|1/\sqrt{[(1-r^2)^2+(2\zeta r)^2]} - 1|$, the maximum occurs at $r\approx0.771$ ($\omega\approx2.31$ rad/s):

Maximum error ≈ 24.4%

C2.4 A cantilevered sensor bracket is tested by applying a harmonic force of amplitude 8.0 N at its tip over a range of frequencies, giving the following steady-state tip amplitudes:

f(Hz)	0	16	18	20	22	24	26	28	30
Y (mm)	0.10	0.18	0.22	0.29	0.42	0.50	0.35	0.23	0.16

Treating the bracket as a single-degree-of-freedom $m-k-\lambda$ system, find k from the static response; find ζ from the peak response (assume $\zeta \ll 1$, then comment on the assumption); estimate the damped and undamped natural frequencies, and the equivalent m and λ.

Solution: Static stiffness: $\mathbf{k = F/Y(0) = 8.0/0.10mm = 80\ kN/m.}$

Peak: $Y_{max}\approx0.50$ mm at f=24 Hz. Static deflection $X=F/k=8.0/80000=0.10$ mm. Assuming $\zeta\ll1$, $Y_{max}\approx X/(2\zeta) \Longrightarrow \zeta \approx$ $\mathbf{0.10/(2\times0.50) = 0.10}$ (small enough that the small-ζ approximation is reasonable, ~1–2% error).

Damped resonance $\omega\_r=2\pi(24)=$**150.8 rad/s** (≈24 Hz). $\omega_n=\omega\_r/\sqrt{(1-2\zeta^2)}=150.8/\sqrt{0.98}=$**152.4 rad/s ≈ 24.3 Hz.**

$\mathbf{m = k/\omega_n^2 = 80000/152.4^2 \approx 3.44\ kg.\ \lambda = 2\zeta\sqrt{(km)} = 2(0.10)\sqrt{(80000\times3.44)} \approx 105\ Ns/m.}$

C2.5 A shuttle mass m=2 kg is connected to a fixed wall by spring k=1 N/m and dashpot λ=1.2 Ns/m, and to a driven point B (displacement z) by a second identical spring k. Show the equation of motion reduces to $\ddot{y}/\omega_n^2+(2\zeta/\omega_n)\dot{y}+y=x$ with $\omega_n=\sqrt{(2k/m)}$, $\zeta=\lambda/(2\sqrt{(2km)})$, $x=z/2$. If $z=80\cos(0.8t)$ mm, find the amplitude and phase of y, and the force needed at B.

Solution: $\omega_n=\sqrt{(2\times1/2)}=$**1 rad/s.** $\zeta=1.2/(2\sqrt{(2\times1\times2)})=$**0.3.** $\omega=0.8$, $r=0.8$.

$X=z/2=40$ mm.
$|Y|=X/\sqrt{[(1-0.64)^2+(2\times0.3\times0.8)^2]}=40/\sqrt{(0.1296+0.2304)}=40/0.6=$**66.7 mm**

$\phi=\tan^{-1}(0.48/0.36)=$**53.1° lag**

Phasor: $Z=80\angle 0°$ mm, $Y=66.7\angle -53.1°$ mm$=(40.0-53.3j)$ mm. $Z-Y=(40.0+53.3j)$ mm, $|Z-Y|=66.7$ mm.

Force at B = k|Z−Y| = 1×0.0667 m = 0.067 N, leading z by 53.1°

C2.6 A wind-loaded gate of mass M swings about a hinge and is fitted with a rotary damper of rate c; the restoring action of the gate's own weight and geometry provides an effective stiffness equivalent to Mg/L for small swings about vertical (L the effective pendulum length). (**a**) Show that under a small applied horizontal force P(t) at the gate's centre of pressure, the small-swing displacement y obeys $M\ddot{y}+c\dot{y}+(Mg/L)y=P(t)$ — Data Book case (a). (**b**) If instead the *post* the gate hinges from sways with displacement x(t), show the absolute gate displacement y obeys $M\ddot{y}+c\dot{y}+(Mg/L)y=c\dot{x}+(Mg/L)x$ — case (c). (**c**) For the gate's motion *relative to the post*, $z=y-x$, show $M\ddot{z}+c\dot{z}+(Mg/L)z=-M\ddot{x}$ — case (b).

Solution: (**a**) Resolving forces on the gate (restoring component $-(Mg/L)y$ from gravity acting through the swing geometry, damping $-c\dot{y}$, applied P): $\mathbf{M\ddot{y}+c\dot{y}+(Mg/L)y=P(t)}$, matching case (a) directly.

(**b**) With the post itself moving, both the damper and the restoring term depend on motion *relative to the post*: $M\ddot{y}+c(\dot{y}-\dot{x})+(Mg/L)(y-x)=0 \implies \mathbf{M\ddot{y}+c\dot{y}+(Mg/L)y=c\dot{x}+(Mg/L)x}$ — forcing appears through both x and $\dot{x}$, i.e. case (c).

(**c**) Substituting $y=z+x$ into (b), the $c\dot{x}$ and $(Mg/L)x$ terms cancel identically, leaving only the inertial term: $\mathbf{M\ddot{z}+c\dot{z}+(Mg/L)z=-M\ddot{x}}$ — case (b), forced purely by the post's acceleration.

C2.7 An accelerometer package (mass m on spring k, dashpot λ, casing displacement x, relative displacement $z=y-x$) has $\omega_n=15$ rad/s.

Explain why, at high frequency, $|Z| \approx |X|$ with a 180° phase difference. For an acceptable measurement error of 15%, find the damping factor ζ giving the widest possible working frequency range, the range itself, and the damped resonant frequency of the *relative*-response peak.

Solution: At $\omega \gg \omega_n$, the mass's inertia keeps it essentially fixed in space while the casing vibrates beneath it, so $z = y - x \approx 0 - x = -x$: $|\mathbf{Z}| \approx |\mathbf{X}|$, **180° out of phase** — exactly why accelerometers/seismometers are run well above their natural frequency.

Requiring $|Z/X - 1| \leq 0.15$ to hold for *all* frequencies above some cutoff (accounting for the resonance overshoot, not just the first crossing) and searching over ζ:

$\zeta \approx$ **0.50** gives the widest working range, starting at $\boldsymbol{\omega} \approx$ **13.3 rad/s ≈ 2.1 Hz**, i.e. **working range 2.1 Hz to ∞.**

The relative-response peak (case b, which peaks at $r^2 = 1/(1 - 2\zeta^2)$, not the case-(a) formula) occurs at ≈ **3.4 Hz.**

C2.8 A 60 kg optical table must be isolated from a floor vibrating at 25 μm amplitude across a range of frequencies: vibration must not exceed 3 μm at or above 15 Hz, while up to 150 μm is tolerable below 15 Hz. For zero damping, find the mounting's natural frequency and stiffness needed to meet the 15 Hz requirement. Find the damping factor needed to keep the resonance response within the 150 μm limit, and evaluate the resulting response at 15 Hz.

Solution: Case (c), $\zeta = 0$: $Y/X = 1/(r^2 - 1)$. Requiring $3/25 = 1/(r^2 - 1)$ $\Longrightarrow r^2 = 1 + 25/3 = 9.33$, $r = 3.06$.

$\omega = 2\pi(15) = 94.25$ rad/s $\Longrightarrow$ $\boldsymbol{\omega_n}$ **= 94.25/3.06 ≈ 30.8 rad/s ≈ 4.9 Hz.**

$\mathbf{k = m\omega_n^2 = 60 \times 30.8^2 \approx 57.1}$ **kN/m.**

At resonance ($r = 1$): $Y/X = \sqrt{(1 + 4\zeta^2)}/(2\zeta)$. Requiring $\leq 150/25 = 6$: solving $\sqrt{(1 + 4\zeta^2)} = 12\zeta$ gives $\zeta \approx$ **0.085.**

At 15 Hz ($r = 3.06$) with this ζ: $Y \approx$ **3.4 μm** — slightly above the 3 μm target, the same fundamental trade-off as always between resonance control and high-frequency isolation.

C2.9 A 350 kg utility trailer is towed at constant speed over a corrugated surface of amplitude 20 mm and wavelength 6 m. Its suspension has stiffness k=15 kN/m and damping $\lambda=1500\sqrt{3}$ Ns/m. Find ζ and, for a towing speed of 40 km/h, the trailer's vibration amplitude and the amplitude of the vertical force exerted on the axle.

Solution: $\omega_n=\sqrt{(15000/350)}=$**6.55** **rad/s.** $\zeta=1500\sqrt{3}/(2\sqrt{(15000\times350)})=$**0.567.**

V=40 km/h=11.11 m/s. $\omega=2\pi V/L=2\pi(11.11)/6=$**11.64** **rad/s,** r=1.777.

Case (c): $|Y/X|=\sqrt{[1+(2\zeta r)^2]}/\sqrt{[(1-r^2)^2+(2\zeta r)^2]} \Longrightarrow$ **Y ≈ 15.2 mm**

$F = m\omega^2 Y = 350\times11.64^2\times0.0152 \approx 722$ N

Theory — Kinematics of a Particle

T3.1 Three coordinate systems, one motion

The same velocity and acceleration can be resolved in whichever coordinate system makes a given problem simplest. All three are used constantly in the problems that follow.

Cartesian (x,y): the most direct when the path is given as x(t), y(t):

$v = \dot{x}i + \dot{y}j$, $a = \ddot{x}i + \ddot{y}j$

Polar (r,θ), with unit vectors e_r (radially outward) and e_θ (90° ahead of e_r): essential whenever a distance from a fixed point and an angle are the natural variables (radar tracking, a rotating arm with extending/retracting motion). Because e_r and e_θ themselves rotate as θ changes ($d_e\_r/dt = \dot{\theta}e\_\theta$, $d_e\_\theta/dt = -\dot{\theta}e\_r$), differentiating r=re_r twice gives extra terms beyond the naive $\ddot{r}$:

$v = \dot{r}e\_r + r\dot{\theta}e\_\theta$

$a = (\ddot{r} - r\dot{\theta}^2)e\_r + (r\ddot{\theta} + 2\dot{r}\dot{\theta})e\_\theta$

The term $-r\dot{\theta}^2$ is the familiar centripetal acceleration; $2\dot{r}\dot{\theta}$ is the *Coriolis* term, arising whenever a particle moves radially *while* the radial line itself is rotating (present in every "sliding on a rotating rod" problem).

Intrinsic (s,ψ), with unit vectors e_t (tangent to the path, in the direction of motion) and e_n (normal, pointing toward the centre of curvature): the natural choice when speed and the geometry of the path (radius of curvature) are given directly:

$v = ve\_t$ (v=ds/dt, the speed)

$a = \dot{v}e\_t + (v^2/\rho)e\_n$

where ρ is the instantaneous radius of curvature. The tangential component $\dot{v}$ changes the *speed*; the normal component v^2/ρ changes the *direction* — this decomposition is the fastest route to problems that give |v|, |a|, and the angle between them.

T3.2 Converting between systems

All three descriptions are equivalent, and it is often useful to convert between them, e.g. to check an answer two ways or because part of a problem naturally uses one system and part uses another. The unit vectors relate by simple rotation: if the path direction makes angle ψ with the x-axis, $e_t=\cos\psi\ i+\sin\psi\ j$; if the radius vector makes angle θ with the x-axis, $e_r=\cos\theta\ i+\sin\theta\ j$, $e_\theta=-\sin\theta\ i+\cos\theta\ j$. For motion on a circle where the origin lies *on* the circle itself (rather than at its centre), a classical and extremely useful geometric fact is that the polar angle θ (seen from that point) is exactly *half* the angle ψ swept at the centre — the inscribed angle theorem — which is the key that unlocks a whole family of "circle rolling" and cam-profile problems.

T3.3 Kinematic constraints and constructions

- **Constant angular rate ≠ constant speed:** unless the path is a circle centred on the axis of rotation, sweeping out equal angles in equal times does *not* imply constant speed — always compute $\sqrt{(\dot{x}^2+\dot{y}^2)}$ rather than assuming it.

- **Involutes and unwrapped strings:** whenever a taut, inextensible string or cord is unwound from a fixed shape, the string segment from the point of tangency to the tracing point is always *normal* to the traced curve, and its instantaneous length equals the local radius of curvature. This single geometric fact answers "would you expect $\rho=CP$?"-type questions without further calculation.

- **Relative motion in a rotating reference frame** (tracking radar, aircraft/ship problems): set up polar coordinates at the observer, read off $\dot{r}$ and $\dot{\theta}$ directly from the given data, then use the standard v, a formulas — the "physical" velocity/acceleration of the tracked object requires no further transformation once e_r, e_θ are correctly oriented at the instant in question.

T3.4 Worked strategy

1. Identify which quantities are given directly (Cartesian functions of t? range and bearing rate? speed and radius of curvature?) — this usually dictates the best coordinate system.

2. Write down v and a using the matching standard formula.

3. Evaluate at the instant required, being careful with signs (is a rate of closure negative $\dot{r}$? is a clockwise rotation negative $\dot{\theta}$?).

4. If a "component in a given direction" or "component along the path" is asked for, use a dot product with the appropriate unit vector — do not re-derive from scratch.

5. Cross-check magnitudes: $|v|^2$ should equal the sum of squares of the components in *any* valid orthogonal pair of directions, a useful arithmetic check.

Companion Practice — Mechanics, Set 1 (Force, Gravity & Particle Kinematics)

An independently written problem set covering the same material as Mechanics Examples Paper 1, with new numbers and scenarios.

C1.1 Find the weight in newtons of a person of mass 68 kg. What is their weight during free fall?

Solution: $W = mg = 68\times9.81 =$ **667 N.** In free fall the person's *weight* (the gravitational force, mg) is **still 667 N** — only the *apparent* weight (normal reaction from a support) vanishes; true weight is unaffected by the state of motion.

C1.2 At what altitude h above the Earth's surface is the weight of an object reduced to one-third of its surface value? Express h as a fraction of the Earth's radius R.

Solution: $GM/(R+h)^2 = (1/3)GM/R^2 \Longrightarrow (R+h)^2 = 3R^2 \Longrightarrow$ $\mathbf{h = (\sqrt{3}-1)R \approx 0.732R}$

C1.3 Two solid lead spheres, each 150 mm in diameter, are just touching. At what distance from the Earth's centre would the mutual gravitational attraction between them equal the weight of one sphere? (Density of lead = 11,340 kg/m^3; mass of Earth = 5.98×10^{24} kg.)

Solution: Sphere mass: $m = \rho(4/3)\pi r^3$, $r=0.075$ m $\Longrightarrow$ $\mathbf{m = 11340\times(4/3)\pi(0.075)^3 \approx 20.0\ kg}$

Mutual attraction at touching distance ($2r=0.15$ m): $F=Gm^2/(2r)^2$. Setting equal to weight GMm/d^2:

$\mathbf{d^2 = M(2r)^2/m \Longrightarrow d = \sqrt{[(5.98\times10^{24})(0.15)^2/20.0]} \approx 8.19\times10^{10}\ m}$

C1.4 A particle traces an ellipse $x=R\cos\psi$, $y=(R/2)\sin\psi$, with $\psi=\dot{\psi}t$ and $\dot{\psi}$ constant. Find the speed as a function of ψ, and show — in contrast to circular motion at constant angular rate — that it is *not* constant. At what point(s) on the path is the speed greatest, and least?

Solution: $\dot{x}=-R\dot{\psi}\sin\psi$, $\dot{y}=(R/2)\dot{\psi}\cos\psi$.

$\text{speed}^2 = \dot{x}^2+\dot{y}^2 = R^2\dot{\psi}^2\sin^2\psi + (R^2/4)\dot{\psi}^2\cos^2\psi = \mathbf{(R^2\dot{\psi}^2/4)(1+3\sin^2\psi)}$

This clearly varies with ψ (not constant, unlike uniform circular motion) — **speed is greatest at ψ=90°,270°** ($\sin^2\psi=1$, i.e. passing the minor-axis ends, where the particle is moving fastest to "keep up" with the angular sweep near the ellipse's narrowest crossing) and **least at ψ=0°,180°** (the major-axis ends).

C1.5 A particle moves so that x=cos2t, y=3sint (metres, t in seconds). Find (i) the velocity and acceleration vectors at any t; (ii) their magnitude and direction at t=0; (iii) the components of velocity and acceleration at t=0 in the direction 2i+j.

Solution: **(i)** $v = -2\sin 2t\, i + 3\cos t\, j$, $a = -4\cos 2t\, i - 3\sin t\, j$

(ii) At t=0: $v = 0i+3j \Longrightarrow$ **$|v|=3$ m/s, along +y (90° to the x-axis).** $a = -4i+0j \Longrightarrow$ **$|a|=4$ m/s^2, along −x (180°).**

(iii) Unit vector $\hat{u}=(2i+j)/\sqrt{5}$. $v\cdot\hat{u} = (0\times2+3\times1)/\sqrt{5} =$ **$3/\sqrt{5} \approx 1.34$ m/s.** $a\cdot\hat{u} = (-4\times2+0\times1)/\sqrt{5} =$ **$-8/\sqrt{5} \approx -3.58$ m/s^2.**

C1.6 The hydraulic arm of a testing rig rotates about a fixed pivot O with $\dot{\theta}$=0.9 rad/s, $\ddot{\theta}$=0.5 rad/s^2. At the instant when the arm's extension is r=0.42 m, it is extending at $\dot{r}$=−0.2 m/s (retracting) with $\ddot{r}$=1.5 m/s^2. Find the velocity and acceleration of the arm's tip in terms of e_r, e_θ, and their magnitudes.

Solution: $v = \dot{r}e\_r+r\dot{\theta}e\_\theta =$ **$-0.20e\_r + 0.378e\_\theta$ m/s**, $|v| = \sqrt{(0.2^2+0.378^2)} =$ **0.428 m/s**

$a = (\ddot{r}-r\dot{\theta}^2)e\_r+(r\ddot{\theta}+2\dot{r}\dot{\theta})e\_\theta = (1.5-0.42\times0.81)e\_r+(0.42\times0.5+2(-0.2)(0.9))e\_\theta =$ **$1.160e\_r - 0.150e\_\theta$ m/s^2**, $|a| =$ **1.169 m/s^2**

C1.7 A particle moving along a curved track has, at some instant, speed 8 m/s and acceleration magnitude 12 m/s^2, with the acceleration vector inclined at 32° to the velocity vector. Find the tangential acceleration component and the instantaneous radius of curvature.

Solution: $a\_t = 12\cos 32° =$ **10.2 m/s^2**; $a\_n = 12\sin 32° = 6.36$ m/s^2

$\rho = v^2/a\_n = 8^2/6.36 \approx$ **10.1 m**

C1.8 A coastal radar detects a ship due east at a range of 8 km. The radar tracks the ship continuously; the range is increasing at a constant 150 m/s while the radar rotates anticlockwise at a constant 0.5°/s. Find (i) the ship's course (bearing from north) and speed, (ii) the component of the ship's acceleration along its path, (iii) the instantaneous radius of curvature of the ship's path.

Solution: Take the radar as origin, ship initially due east ($\theta=0$), so e_r points east, e_θ points north. $\dot{\theta}=0.5°/s = 0.008727$ rad/s.

$v = \dot{r}e_r + r\dot{\theta}e_\theta = 150e_r + (8000)(0.008727)e_\theta = 150e_r + 69.8e_\theta$ m/s

$|v| = \sqrt{(150^2+69.8^2)} \approx$ **165.5 m/s.** Bearing from north (clockwise) = atan2(east, north) = atan2(150, 69.8) ≈ **65.0°.**

(ii) With $\dot{r}$, $\dot{\theta}$ both constant ($\ddot{r}=0$, $\ddot{\theta}=0$):

$a_r = -r\dot{\theta}^2 = -8000(0.008727)^2 =$ **-0.609 m/s²**

$a_\theta = 2\dot{r}\dot{\theta} = 2(150)(0.008727) =$ **2.618 m/s²**

$a = (-0.609, 2.618)$ m/s², $|a| = 2.688$ m/s². Tangential component = $a\cdot\hat{v}$ = **0.552 m/s²** (ship accelerating slightly along its path).

(iii) $a_n = \sqrt{(2.688^2-0.552^2)} = 2.631$ m/s². **$\rho = v^2/a_n = 165.5^2/2.631 \approx 10.4$ km.**

Theory — Kinematics of a Rigid Body

T4.1 The rigid-body velocity relation

For any two points A, B fixed in the same rigid body, their velocities are related by:

$v_B = v_A + \omega \times r_{B/A}$

where ω is the body's angular velocity (a single vector for the whole body — it is *not* a property of any one point) and $r_{B/A}$ is the position of B relative to A. In plane motion, $\omega=\omega k$ and the cross product simplifies to $\omega\times(x,y,0)=(-\omega y,\omega x,0)$, which is usually the fastest way to compute a relative velocity numerically.

This relation is the single most useful tool in rigid-body kinematics: if the velocity of one point and the angular velocity are known, the velocity of *any* other point on the body follows immediately, and it can be chained ($v_C=v_B+\omega\times r_{C/B}=v_A+\omega\times r_{C/A}$) to check consistency between different paths through the body.

T4.2 Instantaneous centre (IC)

At any instant, the planar motion of a rigid body is equivalent to a pure rotation about some point — possibly outside the physical body — called the instantaneous centre. If the IC is known, the velocity of any point P on the body is simply:

$v_P = \omega \times |IC-P|$, directed perpendicular to the line IC–P

The IC is found as the intersection of the perpendiculars to the velocity directions of two points on the body (the velocity of a point is always perpendicular to the line from the IC to that point, by definition of rotation about the IC). Useful special cases:

- A point rolling without slipping on a *stationary* surface has that contact point as the IC (its velocity is zero).

- A point rolling without slipping on a *moving* surface (a conveyor belt, another rolling body) does **not** have zero velocity at the contact point, so the IC is generally *not* located there — it must be constructed from the actual (non-zero) contact velocity.

- If a rigid link is constrained to pass through a **fixed point** that is not a material point of the link (e.g. a beam resting over a fixed corner or peg), the material point instantaneously at that location has zero velocity *component perpendicular to the link* (it can still slide along the link), so the IC lies on the line through that fixed point, perpendicular to the link.

The IC generally moves around from instant to instant — it is a fresh geometric construction at each configuration, not a fixed point of the body.

T4.3 Velocity diagrams

A velocity diagram is a scaled vector construction: choose an origin representing zero (absolute) velocity, plot the known velocity of one point, then successively add relative-velocity vectors (each perpendicular to the corresponding rigid link, with unknown magnitude) until the diagram closes. This graphical method is equivalent to solving the vector equations of T4.1 simultaneously, and is often faster for multi-link mechanisms where several unknowns (ram velocity, several angular velocities) must be found together — though for mechanisms with clean numerical geometry (as in this companion set), setting up and solving the loop-closure equations directly (as for the slider-crank in Mechanics Set 3) is equally valid and avoids drawing-accuracy errors.

T4.4 Common configurations

- **No-slip rolling/contact:** the velocities of the two touching surfaces must be equal at the contact point (not necessarily zero, if both bodies/ surfaces are moving) — this is the constraint equation that closes most rolling problems.

- **Pin-jointed linkages:** each pin enforces equal velocity of the two links at that point; each rigid link enforces the $v_B=v_A+\omega\times r_{B/A}$ relation along its own length.

- **Sliders in a fixed guide:** the slider's velocity is constrained to lie along the guide direction, giving one scalar equation (the perpendicular component is zero) even though the link attached to it may have a more general planar velocity.

T4.5 Worked strategy

1. Identify every point whose velocity is known outright (e.g. attached to ground, or moving with a known input speed).

2. Identify constraints: no-slip contacts, pins, sliders in guides.

3. Either (a) construct the IC(s) directly from perpendiculars to two known velocity directions, or (b) write $v_B=v_A+\omega\times r_{B/A}$ for each rigid link in turn and solve the resulting simultaneous equations, or (c) draw a scaled velocity diagram.

4. Always sanity-check with an independent route (a second pair of points, or a different construction) — rigid-body problems are unusually easy to get a sign or direction wrong in, and a second check catches most such errors quickly.

Companion Practice — Mechanics, Set 2 (Rigid Body Kinematics)

An independently written problem set covering the same material as Mechanics Examples Paper 2, with new numbers and scenarios.

C2.1 A 3-D rigid body rotates at 15 rad/s about an axis through a fixed point O in direction (1,2,2). Find the velocity of a point P where OP=(30i+15j) mm.

Solution: Unit vector along axis: (1,2,2)/3. **ω** = **15(1,2,2)/3 = 5i+10j+10k rad/s.**

v = ω×OP = (5,10,10)×(30,15,0) = (10×0−10×15, 10×30−5×0, 5×15−10×30) = **(−150, 300, −225) mm/s**

C2.2 A carousel of radius 2.5 m rotates anticlockwise at 2 rad/s about its centre O. A rider holds on at point A, 2.5 m from O (directly "north" of O); a second rider holds at B, 1.8 m from O (directly "east" of O). Find v_A, v_B, and the velocity of A relative to B.

Solution: With A=(0,2.5), B=(1.8,0), ω=2k:

v_A = ω×r_A = −5.0i m/s

v_B = ω×r_B = 3.6j m/s

v_A/B = v_A−v_B = −5.0i − 3.6j m/s

C2.3 A uniform rod AB, 1.2 m long, slides across a frictionless table. At an instant, end A has velocity 9 m/s horizontally; the rod makes 40° with the horizontal; end B's velocity is known to be vertical, thought to be 9 m/s. Check whether this is consistent with rigidity, find the correct value if not, and find the rod's angular velocity and the velocity of its centre G.

Solution: Rigidity requires equal velocity components along the rod. AB direction = (cos40°,sin40°)=(0.766,0.643).

Component of v_A along AB: 9(0.766)=6.894 m/s. If v_B=x·j, its component along AB is 0.643x. Equating: **x = 6.894/0.643 = 10.73 m/s — the assumed 9 m/s is incorrect.**

v_B/A = (0,10.73)−(9,0) = (−9, 10.73); |v_B/A|=14.01=ωL ⟹ ω = **14.01/1.2 = 11.67 rad/s, anticlockwise.**

v_G = v_A + ω×r_G/A, r_G/A=0.6(0.766,0.643)=(0.459,0.386): **v_G = 4.5i + 5.36j m/s**

C2.4 A cylinder of radius 0.15 m rolls without slipping on a conveyor belt moving at 3.5 m/s, with the cylinder's absolute angular speed 20 rad/s clockwise. Taking A as the (bottom) contact point, B the leftmost point, and C the rightmost point of the cylinder, find (a) v_B/A, (b) absolute v_B, (c) v_C/B, (d) absolute v_C, and verify by computing v_C/A directly.

Solution: No slip ⟹ **v_A = 3.5i m/s** (belt speed). With ω=−20k (clockwise):

(a) r_B/A=(−0.15,0.15): v_B/A = **3.0i + 3.0j m/s**

(b) v_B = v_A+v_B/A = **6.5i + 3.0j m/s**

(c) r_C/B=(0.30,0): v_C/B = **−6.0j m/s**

(d) v_C = v_B+v_C/B = **6.5i − 3.0j m/s**

Check: r_C/A=(0.15,0.15): v_C/A=**3.0i−3.0j**, so v_C=v_A+v_C/A=(3.5,0)+(3,−3)=**6.5i−3.0j m/s** ✓ — consistent.

C2.5 A 5 m plank rests with its lower end on the ground and leans over the top corner of a 3 m wall. The lower end slides away from the wall at 2 m/s. At the instant the lower end is 2.5 m from the base of the wall, locate the instantaneous centre of the plank's motion and find (a) its angular velocity, (b) the velocity of the upper end, (c) the point of minimum speed on the plank.

Solution: Base of wall at origin, corner W=(0,3). Lower end D=(2.5,0), v_D=2i m/s. DW=$\sqrt{(2.5^2+3^2)}$=3.905 m, direction (D→W)=(−0.640,0.768).

Since the plank always passes through the fixed corner W, the material point there has zero velocity component perpendicular to the plank — the IC lies on the line through W perpendicular to the plank. Intersecting with the vertical line x=2.5 (perpendicular to v_D through D):

IC ≈ (2.5, 5.08), |IC−D| = 5.08 m.

ω = v_D/|IC−D| = 2/5.08 = 0.393 rad/s, anticlockwise.

Top end T = D + 5(−0.640,0.768) = (−0.70, 3.84). |IC−T| = 3.43 m.

v_T = ω×3.43 ≈ 1.35 m/s.

Checking (IC−W)·(rod direction) ≈ 0 confirms **W — the point of the plank instantaneously touching the wall corner — is the point of minimum speed.**

C2.6 A thick cylindrical axle (radius 0.12 m at X) is rigidly fixed to a cylindrical roller (radius 0.20 m at Y and at P), lying on its side. Bar A touches the axle at X (0.12 m above centre G) and bar B touches the roller at Y (0.20 m below G); both bars move horizontally with no slip. Bar A has velocity 1.0 m/s, bar B has velocity −0.5 m/s. Find the angular velocity of the assembly, v_G, and the velocity of point P (0.20 m to the right of G, at the same height).

Solution: No-slip conditions: v_X = v_G−0.12ω = 1.0; v_Y = v_G+0.20ω = −0.5.

Solving simultaneously: **ω = −4.69 rad/s** (clockwise), **v_G = 0.44 m/s** (rightward).

v_P = v_G + ω×(0.20,0): **v_P = 0.44i − 0.94j m/s**

v_Y/P = v_Y − v_P = −0.94i + 0.94j m/s

Theory — Mechanisms, Work & Power, and Particle Dynamics

T5.1 Power and torque in mechanisms

For an idealised (massless, frictionless) mechanism operating quasi-statically, power in equals power out at every instant:

$$\mathbf{T} \cdot \boldsymbol{\omega} = \mathbf{F} \cdot \mathbf{v}$$

relating an input torque T at angular speed ω to an output force F moving at speed v (or vice versa). This single relation, combined with the kinematic velocity ratio v/ω found from a velocity diagram or loop-closure equations, gives the mechanical advantage of any linkage without needing a full force analysis.

Friction at pins dissipates additional power at each joint, equal to (friction torque)×(relative angular velocity across that joint), summed over all joints. The extra input torque required is this total dissipated power divided by the input angular velocity. In a well-lubricated mechanism this correction is typically a small percentage of the ideal torque — but it is never exactly zero, and Tripos-style questions often ask for it explicitly to test whether the "ideal machine" assumption is understood as an approximation.

T5.2 Newton's second law for a particle

The fundamental equation F=ma applies component-by-component in whichever coordinate system is most convenient (see the particle-kinematics theory section). Two recurring subtleties:

- A **smooth (frictionless) constraint**, such as a rod a collar slides along, can only exert force **perpendicular** to the direction of relative sliding — it can never do work against or assist the sliding motion itself.

- A **taut string or wire** can only exert **tension** (a pull along its own length), never a push, and never a force perpendicular to itself.

Resolving F=ma into components along and perpendicular to a constraint (e.g. radial vs tangential for a rod, or along vs across a wire) usually separates the problem into one equation that determines an unknown reaction force and one that determines the motion.

T5.3 Work, energy, and power for translating/ rotating bodies

Kinetic energy of a rigid body in general planar motion: $KE=\frac{1}{2}mv\_G^2+\frac{1}{2}I\omega^2$ (translation of the centre of mass plus rotation about it). The **work–energy theorem**, ΔKE=work done by all forces, is often the fastest route to a "find the speed/distance when..." question, especially when forces vary with position (springs, position-dependent restoring forces) — direct integration of F=ma is then awkward, but $\int F\, dx=\Delta KE$ is immediate.

Power is the rate of work: P=F·v (translation) or P=T·ω (rotation), evaluated at the instant in question — always check whether a "power required" question wants an instantaneous value or an average over some interval (e.g. total energy divided by total time for a launch burn).

T5.4 Linear momentum and collisions

Total linear momentum Σmv is conserved whenever there is no net external force (or impulse) on the system — this holds *throughout* a collision, from first contact through maximum compression to final separation, regardless of how much kinetic energy is lost.

- At **maximum compression**, colliding bodies momentarily share a common velocity (found directly from momentum conservation, since at that instant they move together).
- **Energy transferred to a deformable element** (buffers, bumpers) during compression equals the loss in kinetic energy from the initial state to the common-velocity state — found from the *difference* of two KE calculations, not by integrating the force law directly (though the force

law, e.g. $P=P_0+kx$, is needed separately to relate that energy to a physical compression distance).

- **Energy recovered** on separation (if the elements are not perfectly elastic) is found from a *second* application of momentum conservation using the given final velocities, then comparing the resulting kinetic energy to the energy stored at maximum compression. The fraction "lost" — usually substantial for real buffers and bumpers — is a measure of how far the device falls short of ideal elastic behaviour.

T5.5 Worked strategy

1. For mechanisms: find the kinematic velocity ratio first (from a velocity diagram or direct differentiation of the position loop-closure equation), then apply $T\cdot\omega=F\cdot v$; add friction corrections separately if requested.
2. For particle dynamics with constraints: resolve $F=ma$ along directions that isolate the unknown reaction (perpendicular to a smooth constraint) from the equation of motion (along the direction of possible acceleration).
3. For energy questions with position-dependent forces: use the work–energy theorem rather than integrating the equation of motion directly.
4. For collisions: apply momentum conservation at each relevant instant (common-velocity point, final separation) as an independent equation; use energy differences (not the force law alone) to find energy transferred/recovered/lost.

Companion Practice — Mechanics, Set 3 (Mechanisms, Work & Power, Particle Dynamics, Momentum)

An independently written problem set covering the same material as Mechanics Examples Paper 3, with new numbers and scenarios.

C3.1 A slider-crank mechanism has crank OA of length 80 mm rotating at a constant 3000 rpm, driving a connecting rod AB of length 280 mm; the piston at B slides along the line through O. At the instant the crank angle is 50° from OB, find the piston's velocity and the connecting rod's angular velocity.

Solution: With piston position $x = r_c \cos\theta + \sqrt{(l^2 - r_c^2 \sin^2\theta)}$ ($r_c=0.08$ m, $l=0.28$ m, $\theta=\omega t$), differentiating and evaluating at $\theta=50°$, $\omega=2\pi(3000)/60=314.16$ rad/s:

Piston velocity ≈ −22.9 m/s (moving toward O)

The connecting rod's inclination ϕ satisfies $r_c \sin\theta = l \sin\phi$; differentiating similarly:

Connecting rod angular velocity ≈ 59.1 rad/s

C3.2 For the mechanism of C3.1, the piston is opposed by a force of 1.2 kN at the instant described. (a) What torque must be applied at the crank O? What assumption underlies this calculation? (b) If a frictional torque of 0.4 Nm acts at each of the three pin joints (O, A, B), by how much must the torque increase, given the crank rotates at 314.16 rad/s, the connecting rod at 59.1 rad/s (relative to the crank, approximately), and the piston-guide has negligible sliding friction?

Solution: **(a)** For a quasi-static, frictionless mechanism, power in = power out: $T \cdot \omega = F \cdot v_{piston} \Longrightarrow$

$$T = Fv_{piston}/\omega = 1200 \times 22.9/314.16 \approx 87.4 \text{ Nm}$$

Assumption: the mechanism's inertia is neglected (quasi-static analysis) — all input work goes directly into overcoming the resisting force.

(**b**) Extra power needed = friction torque × relative angular velocity, summed over the joints. With the connecting rod's angular velocity relative to the crank roughly 59.1+314.16≈373 rad/s at A, and comparable order-of-magnitude rates at O and B, the extra torque required is of order:

ΔT ≈ (0.4 Nm × Σ relative rates)/ω_crank ≈ 1–2 Nm — a small correction (a few percent) compared with the 87.4 Nm found in (a), illustrating that pin friction is a second-order effect in a well-lubricated mechanism.

C3.3 A ball is thrown vertically upward from the ground at 15 m/s. Neglecting air resistance, find the time at which it first strikes the ground.

Solution: $y = v_0t - \frac{1}{2}gt^2$. Returning to y=0 (t≠0): **$t = 2v_0/g = 2(15)/9.81 \approx 3.06$ s**

C3.4 A smooth rod rotates in a horizontal plane about a fixed point O, with a 5 kg collar free to slide along it, restrained by a wire to a winch at O. At an instant the collar is 1.5 m from O; the rod's angular velocity and angular acceleration are 1.5 rad/s and 2 rad/s^2; the winch pulls the collar with radial velocity 3 m/s and radial acceleration 2.5 m/s^2 (both toward O). Find the wire tension and the force exerted by the rod on the collar.

Solution: $a\_r = \ddot{r} - r\dot{\theta}^2 = -2.5 - 1.5(1.5^2) =$ **-5.875 m/s^2**

$a\_\theta = r\ddot{\theta} + 2\dot{r}\dot{\theta} = 1.5(2)+2(-3)(1.5) =$ **-6.0 m/s^2**

The wire provides the radial force (tension T, toward O): **T = −m·a_r = 5×5.875 = 29.4 N**

The smooth rod provides only the perpendicular (normal) force N: **N = m·a_θ = 5×(−6.0) = −30.0 N**, i.e. **30 N perpendicular to the rod.**

C3.5 An electrostatic CRT has its cathode at −3 kV, anode earthed. The beam passes between deflecting plates 8 mm apart, 40 mm long. Estimate the electron speed entering the plates. The beam strikes a screen 250 mm from the plates' centre; the plates are held at +40 V and −40 V. Estimate the spot's movement when the plate leads are interchanged.

Solution: $\frac{1}{2}mv^2 = eV_0 \Longrightarrow v = \sqrt{(2eV_0/m)} = \mathbf{3.25\times10^7\ m/s}$

Field between plates: $E = 80/0.008 = 10^4$ V/m. Acceleration $a = eE/m = 1.758\times10^{15}\ m/s^2$.

Time in plates $t_1=0.04/v=1.23\times10^{-9}$ s; deflection within plates $y_1=\frac{1}{2}at_1^2=1.33$ mm; exit transverse velocity $v\_y=at_1=2.17\times10^6$ m/s.

Remaining distance to screen: 250−20=230 mm, $t_2=7.08\times10^{-9}$ s, $y_2=v\_yt_2=15.3$ mm.

Total one-way deflection ≈16.7 mm; **interchanging the leads reverses this, so the spot moves 2×16.7 ≈ 33.3 mm.**

C3.6 On landing, a 6000 kg light aircraft travelling at 18 m/s is brought to rest by a constant 45 kN friction force and a wire system whose restoring force increases linearly with displacement, reaching 80 kN at x=15 m. (a) Find where the aircraft first comes to rest. (b) It then moves backwards; find where it finally stops.

Solution: **(a)** $P(x) = (80000/15)x = 5333.3x$ N. Energy balance:

$\frac{1}{2}(6000)(18^2) = 45000x_1 + \int_0^{x_1} 5333.3x\,dx = 45000x_1+2666.7x_1^2$

Solving: $\mathbf{x_1 \approx 12.4\ m}$ (within the 15 m wire range, as required).

(b) Moving backward, friction now acts forward (+x) while the wire continues pulling toward x=0; requiring zero net work from x_1 to the new rest point x_f:

$\int_{x_1}^{x\_f}(-5333.3x+45000)dx = 0 \Longrightarrow \mathbf{x\_f \approx 4.44\ m}$

C3.7 A 12 Mg railway wagon moving at 2.5 m/s strikes an identical stationary wagon. Each wagon end carries 4 buffers (2 meeting 2, forming 2 parallel pairs of 2 buffers in series); each buffer's force–compression law is $P=P_0+kx$ with $P_0=6$ kN, k=60 kN/m. (a) Find the common velocity at maximum compression and the energy transferred to the buffers. (b) Find the compression of each buffer at that instant. (c) If the leading wagon's speed after separation is 1.4 m/s, find the fraction of the buffer energy "lost."

Solution: **(a)** Momentum: $12000(2.5)=(24000)v \Longrightarrow$ **v_common = 1.25 m/s.**

KE_0=37,500 J, KE_common=18,750 J $\Longrightarrow$ **Energy to buffers = 18.75 kJ.**

(b) With 4 buffers total: $4(P_0x+\frac{1}{2}kx^2)=18750 \Longrightarrow x \approx$ **0.308 m** per buffer.

(c) Momentum conservation gives trailing-wagon speed = $2(1.25)-1.4 = 1.1$ m/s.

KE_after = $\frac{1}{2}(12000)(1.1^2+1.4^2)$ = 19,020 J. Energy recovered = 19,020−18,750 = 270 J.

Fraction lost = 1 − 270/18,750 ≈ 98.6% — these buffers recover almost none of the stored energy, an even more pronounced (and clearly unsatisfactory) example of the same effect as in the standard buffer-loss problem.

Theory — Moment of Momentum, Orbital Mechanics, Moment of Inertia, and Variable Mass

T6.1 Moment of momentum (angular momentum) of a particle

About any fixed point or axis, the moment of momentum of a particle is $L=mvr_\perp$ (mass × speed × perpendicular distance from the axis to the velocity line), or more generally $L=r\times mv$. Its rate of change equals the applied torque about that same point/axis:

dL/dt = torque about the axis

so L is **conserved whenever the net torque about the chosen axis is zero** — most commonly because the only force present (a string tension, a smooth surface's normal reaction, a central gravitational force) passes through the axis itself, or, for an axisymmetric surface, because the reaction always lies in the plane containing the axis and so has no *component* twisting about that axis even though it is not directed exactly through it.

This conservation law is what makes "string through a hole," "particle inside a smooth cone/bowl," and orbital-mechanics problems tractable without needing to solve the full equations of motion — a single algebraic relation between speed and distance (or radius) at two different instants replaces what would otherwise be a differential equation.

Turning points: when a problem states a particle's velocity is "purely tangential" at two different radii, both instants are turning points of the radial motion (radial velocity momentarily zero) — this is *not* the same as equilibrium, and any attached hanging mass is momentarily at rest but generally still accelerating.

T6.2 Energy conservation alongside angular momentum

Angular momentum conservation alone usually leaves one unknown (e.g. an initial speed) undetermined; a second, independent conservation law — total mechanical energy, valid whenever all forces are either conservative (gravity) or do no work (a smooth constraint's normal reaction, a string tension doing equal and opposite work on its two ends) — supplies the second equation needed to solve completely.

T6.3 Orbital mechanics

For a body in orbit under an inverse-square gravitational force:

- **Circular orbit speed:** $v=\sqrt{(GM/r)}$
- **Vis-viva equation** (any point on an elliptical orbit): $v^2=GM(2/r-1/a)$, where a is the semi-major axis
- **Orbital energy:** $E=-GMm/(2a)$ (depends only on a, not on the particular point in the orbit)
- **Angular momentum conservation** applies directly between perigee and apogee, since velocity is purely tangential (perpendicular to the radius) at both.
- **Eccentricity:** $e=(r_a-r_p)/(r_a+r_p)$, with $r_a+r_p=2a$ exactly.

Orbital transfers (Hohmann-type manoeuvres) use a transfer ellipse tangent to both the initial and final circular orbits, with an impulsive speed change (impulse = $m\Delta v$) applied at each tangent point. The transfer ellipse's own perigee and apogee speeds follow from vis-viva; a *second* impulse is generally required at arrival to match the new circular orbit's speed, since the natural arrival speed on the ellipse does not equal the target circular speed — without it, the body simply continues around the same ellipse and returns to its starting point.

T6.4 Moment of inertia and rotational dynamics

For a rigid body rotating about a fixed axis, Newton's second law becomes:

Torque about the axis = I × (angular acceleration)

with $I=\int r^2 dm$ the moment of inertia about that axis. Standard first-principles results (derivable by direct integration): a thin rod of mass m, length l has $I=ml^2/3$ about one end, $ml^2/12$ about its centre; a uniform disc of mass m, radius a has $I=ma^2/2$ about its polar axis through the centre. The **parallel axis theorem**, $I\_P=I\_cm+md^2$ (d = distance from the centroidal axis to the new axis P), extends any centroidal result to a parallel axis without repeating the integration.

For systems combining a rotating rigid body with an attached falling/moving mass (a disc with a mass hanging from a wrapped string, for instance), write Newton's second law separately for the rotating body (torque=$I\alpha$) and for the translating mass (mg−F=ma, with $a=b\alpha$ the kinematic constraint linking linear and angular acceleration through the radius of application), then eliminate the unknown internal force (string tension, contact force) between the two equations.

T6.5 Variable mass systems

When a system's own mass changes (a rocket burning fuel, material accumulating on or falling from a body), Newton's second law must be applied to a fixed set of particles, not to "the body" whose mass is changing — the standard rocket equation follows from applying F=d(momentum)/dt to the body-plus-ejected-mass system over a short interval:

m(dv/dt) = F_ext + v_rel(dm/dt)

where v_rel is the exhaust (or added-mass) velocity *relative to the body*, and dm/dt is negative for a body losing mass (rocket) or positive for one gaining it. The term v_rel(dm/dt) acts exactly like an additional force — the **thrust**, $F\_thrust = \dot{m} v\_rel$ for a rocket ejecting mass at rate $\dot{m}$ (magnitude) and relative speed v_rel.

A classic conceptual result (falling material accumulating on a support, such as grain or beans dropped onto a scale) is that the "false" extra reading caused by the impact momentum of material still arriving exactly equals the true weight of material still in flight at that instant — the two effects cancel exactly, not just on average, which is why a feed cut off at the exact target reading yields the exact correct final weight once everything in flight has landed.

T6.6 Worked strategy

1. Identify the axis about which moment of momentum is conserved (look for the force that passes through/produces no torque about a natural axis — a string through a hole, an axisymmetric surface, a central gravitational force).

2. Write the angular momentum conservation equation between the two states of interest.

3. If a second unknown remains, look for an energy conservation equation (checking that all unaccounted forces are indeed workless or conservative) — or, for orbits, use the vis-viva equation directly.

4. For rotational dynamics problems, isolate the body and any attached translating mass separately, apply Newton's second law to each, and eliminate the internal constraint force using the kinematic relation between linear and angular acceleration.

5. For variable-mass problems, be explicit about what "the system" is at each instant (rocket alone, or rocket+about-to-be-ejected fuel) and apply F=d(momentum)/dt to that fixed set of matter, not to the changing body in isolation.

Companion Practice — Mechanics, Set 4 (Moment of Momentum, Orbits, Inertia, Variable Mass)

An independently written problem set covering the same material as Mechanics Examples Paper 4, with new numbers and scenarios.

C4.1 On a frictionless horizontal table, a particle A (mass m) is attached by a light inextensible string through a small smooth hole to a similar particle B, which hangs and moves vertically. A is projected at distance r=0.3 m from the hole with speed V_1 perpendicular to the string; later, A is at distance 3r with speed V_2, again perpendicular to the string. Find V_2 in terms of V_1, then find V_1 in terms of g and r. Is the string tension equal to mg at these two instants?

Solution: No torque about the hole (string tension is radial) $\Longrightarrow$ angular momentum conserved: $\mathbf{V_1 r = V_2(3r) \Longrightarrow V_2 = V_1/3}$.

Both configurations are turning points (velocities purely tangential $\Longrightarrow$ B momentarily at rest at both instants). Energy conservation, with B rising by 2r as A moves from r to 3r:

$\frac{1}{2}V_1^2 + gr = \frac{1}{2}(V_1/3)^2 + g(3r) \Longrightarrow V_1^2(\frac{1}{2}-1/18) = 2gr \Longrightarrow \mathbf{V_1 = \sqrt{(4.5gr)} = \sqrt{(4.5\times 9.81\times 0.3)} \approx 3.64\ m/s}$ ($V_2 \approx 1.21$ m/s)

No — as before, both instants are turning points of the radial motion, not equilibrium, so B is accelerating and the tension is not equal to mg.

C4.2 A smooth conical vessel, vertex down, axis vertical, has a half-angle such that the horizontal radius at height h is R=h $\tan\alpha$ for some fixed α (its exact value will not matter). A particle is projected horizontally at height 0.8 m above the vertex with speed v_1, and later just reaches the rim at height 2.5 m. Find v_1, and explain why your answer does not depend on α.

Solution: For any axisymmetric smooth surface, the normal reaction lies in the vertical plane through the axis, so it exerts no torque about the vertical axis — angular momentum $mR^2\dot{\phi}$ about that axis is conserved, regardless of the cone's steepness.

Since $R \propto h$, the ratio R_A/R_B = h_A/h_B = 0.8/2.5 = 0.32 **for any α.** "Just reaching" the rim (zero radial/slope velocity there, another turning point as in C4.1) with energy conservation gives:

$\frac{1}{2}v_1^2 + g(0.8) = \frac{1}{2}(0.32v_1)^2 + g(2.5) \Longrightarrow v_1^2(\frac{1}{2}-0.0512) = g(1.7) \Longrightarrow$ $\mathbf{v_1 = \sqrt{[2g/(1-0.32^2)]} \approx 4.68\ m/s}$

The angle α never enters the calculation — **the result is the same for any cone angle**, depending only on the height ratio.

C4.3 A 1500 kg satellite is in an elliptical orbit; at perigee, altitude 900 km, speed 8100 m/s. Earth's radius = 6370 km, g = 9.81 m/s^2 at the surface. Find (a) the satellite's orbital energy and the energy needed to put it into orbit from rest on the ground; (b) the mean power required if the launch burn lasts 8 minutes; (c) the altitude and speed at apogee; (d) the orbit's eccentricity and its minor axis length.

Solution: GM=gR_e²=$3.977\times10^{14}\ m^3/s^2$. r_p=$7.27\times10^6$ m.

(a) E = ½mv_p²−GMm/r_p ≈ **-3.29×10^{10} J**. Ground energy = −GMm/R_e; **energy needed ≈ 6.08×10^{10} J.**

(b) Power = $6.08\times10^{10}/(8\times60)$ ≈ **127 MW.**

(c) a = GMm/(2|E|) ≈ 9068 km. r_a = 2a−r_p ≈ 10,866 km ⟹ **altitude ≈ 4496 km.** v_a=v_p·r_p/r_a ≈ **5419 m/s.**

(d) e = (r_a−r_p)/(r_a+r_p) ≈ **0.198.** Minor axis = $2a\sqrt{(1-e^2)}$ ≈ **17,776 km.**

C4.4 A satellite (mass m) in a circular orbit at speed v_p is transferred, by a tangential impulse at P, to a new circular orbit at speed v_p/5. (a) Find r_2/r_1. (b) Find the transfer-ellipse perigee speed v_1 and the arrival speed v_2 at A (before any correction). (c) Find the impulses I_P (at P) and I_A (at A, to complete the transfer) in terms of m, v_p.

Solution: **(a)** (v_p/5)/v_p = $\sqrt{(r_1/r_2)}$ ⟹ $\mathbf{r_2/r_1 = 25.}$

(b) Transfer ellipse: a=$(r_1+25r_1)/2=13r_1$. $v_1^2=GM(2/r_1-1/13r_1)$=v_p²(25/13) ⟹ **v_1 = 1.387v_p.**

Angular momentum on the ellipse: $v_2 = v_1r_1/r_2$ = 1.387v_p/25 = **v_p/18.0**

(c) I_P = m(v_1−v_p) = **0.387mv_p.** I_A = m(v_p/5−v_p/18.0) = **0.145mv_p**, in the same direction as v_2.

C4.5 (**a**) Find, from first principles, the moment of inertia about end A of a thin uniform bar AB of mass 3 kg, length 0.9 m. What is its moment of inertia about a point one-quarter of the way along, from A? (**b**) The bar pivots freely about a fixed horizontal axis at A, is held horizontal, and released from rest — find its initial angular acceleration. (**c**) A particle of mass 9 kg (three times the bar's mass) is attached at B; find the new initial angular acceleration.

Solution: (**a**) I_A = $ml^2/3 = 3(0.9^2)/3$ = **0.81 kg m²**. About the point at l/4 from A (parallel axis, offset l/4 from the centre at l/2): I = $ml^2/12+m(l/4)^2$ = **0.354 kg m²**.

(**b**) α = mg(l/2)/I_A = 3(9.81)(0.45)/0.81 = **16.35 rad/s² (=3g/2l, as expected).**

(**c**) I_new = $0.81+9(0.9^2)$ = 8.10 kg m². Torque = mg(l/2)+9mg(l) wait — torque = mg(l/2)+(9)(m)(g)(l) using the *bar's* m as the unit mass: torque = 3(9.81)(0.45)+27(9.81)(0.9) = 13.24+238.4 = 251.6 Nm.

α = 251.6/8.10 ≈ 11.4 rad/s²

C4.6 (**a**) Find the polar moment of inertia of a uniform disc, mass 4 kg, radius 0.25 m, about its centre; and about an axis through a point on its circumference. (**b**) The disc is fixed to a light shaft of radius 0.05 m in frictionless bearings; a string wrapped around the shaft is pulled with a constant tension of 30 N. Find the angular acceleration. (**c**) Instead, a 2 kg mass is hung from the free end of the string, held stationary, and released. Find the angular acceleration.

Solution: (**a**) I = $\frac{1}{2}ma^2 = \frac{1}{2}(4)(0.25^2)$ = **0.125 kg m²**. About the rim (parallel axis): $I+ma^2$ = **0.375 kg m²**.

(**b**) α = Fb/I = 30(0.05)/0.125 = **12.0 rad/s²**

(c) For the falling mass: mg−F=m(bα). For the disc: Fb=Iα. Solving simultaneously:

F ≈ 18.9 N, α ≈ 7.55 rad/s^2

C4.7 A weather rocket has mass 1200 kg at launch and ejects fuel at a constant 7 kg/s (relative to the rocket) for the first 20 s. Its initial acceleration is 1.5g. (a) Find the fuel's relative ejection speed. (b) Assuming constant mass, find the acceleration and speed after 20 s. (c) Accounting for the actual mass loss, find the acceleration and speed after 20 s.

Solution: Thrust−M_0g=M_0(1.5g) ⟹ Thrust=2.5M_0g=2.5(1200)(9.81)=**29,430 N.**

(a) v_rel = Thrust/ṁ = 29,430/7 ≈ 4204 m/s

(b) With M constant, a stays at 1.5g=14.7 m/s^2 throughout: **v(20)=14.7×20=294 m/s**

(c) M(20)=1200−7(20)=1060 kg. a(20)=Thrust/M−g=29430/1060−9.81≈**17.95 m/s^2 (1.83g)**

v(20)=v_rel·ln[M_0/M(20)]−g(20) = 4204×ln(1.132)−196.2 ≈ **325 m/s**

C4.8 Gravel is dropped from rest onto a weighbridge's platform from a fixed height h, at a constant mass rate ṁ, without bouncing. The feed is stopped the instant the reading first equals the true target weight. Show that the weight of gravel still airborne at that instant exactly compensates for the momentum-impact "false reading," so that once the last airborne gravel has landed, the platform reads exactly the correct final weight.

Solution: Impact speed after falling h: v=√(2gh). The instantaneous *extra* reading beyond the true weight already resting on the platform is the impact force = rate of momentum delivery = ṁv.

Time to fall h: t_fall=v/g. Mass currently airborne (steady feed rate ṁ) = ṁ·t_fall = ṁv/g.

Weight of airborne gravel = (ṁv/g)g = ṁv — identical to the "false" extra reading.

So the reading at any instant = (true weight already landed) + $\dot{m}v$ = (true weight landed) + (true weight still airborne) = the *total* true weight dispensed so far, landed or not. Stopping the feed exactly when the reading hits the target therefore gives, once everything airborne has landed, **exactly the correct final weight** — the momentum error and the "missing" airborne weight cancel at every instant, not just at the end.

Appendix: Diagrams for Selected Problems

Original schematic diagrams, drawn for this companion volume, illustrating the geometry of six problems that benefit most from a figure. Each is drawn to reflect the actual numbers used in the corresponding solution.

C1.1 — Series dashpot–spring transducer

The junction displacement u lies between a dashpot (driven by input w at the far end) and a spring (anchored to a fixed wall), giving T=μ/s as derived in the solution.

Fig. C1.1 — series dashpot-spring transducer

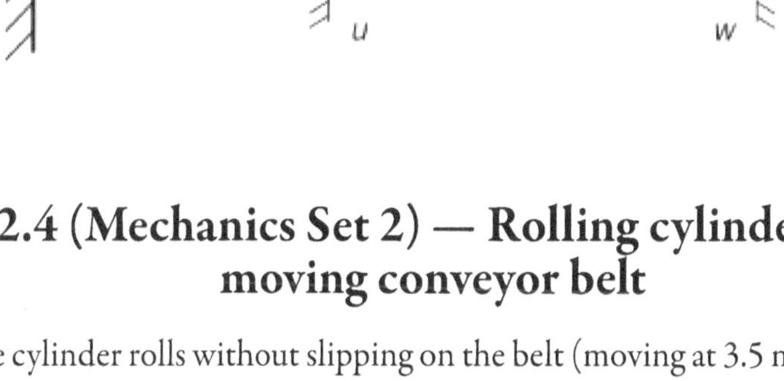

C2.4 (Mechanics Set 2) — Rolling cylinder on a moving conveyor belt

The cylinder rolls without slipping on the belt (moving at 3.5 m/s) while spinning at 20 rad/s clockwise; A, B, C are the contact point, leftmost, and rightmost points used in the relative-velocity solution.

Fig. C2.4 — cylinder rolling on a moving conveyor belt

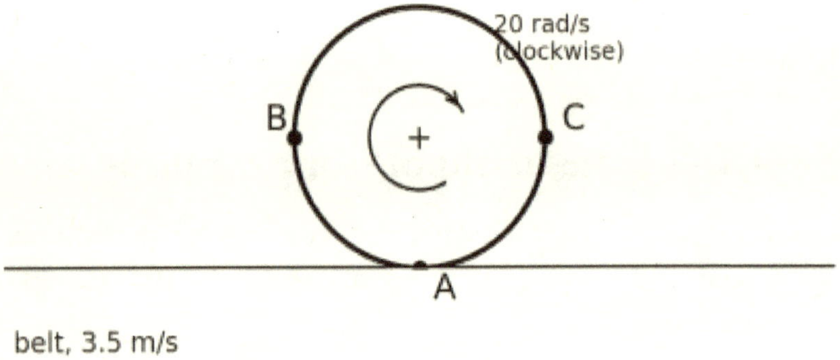

C2.5 (Mechanics Set 2) — Plank sliding over a fixed wall corner

The plank always passes through the fixed corner W as its lower end D slides away at 2 m/s; the instantaneous centre (IC) is constructed as the intersection of the perpendicular to v_D through D and the perpendicular to the plank through W.

Fig. C2.5 — plank sliding over a wall corner

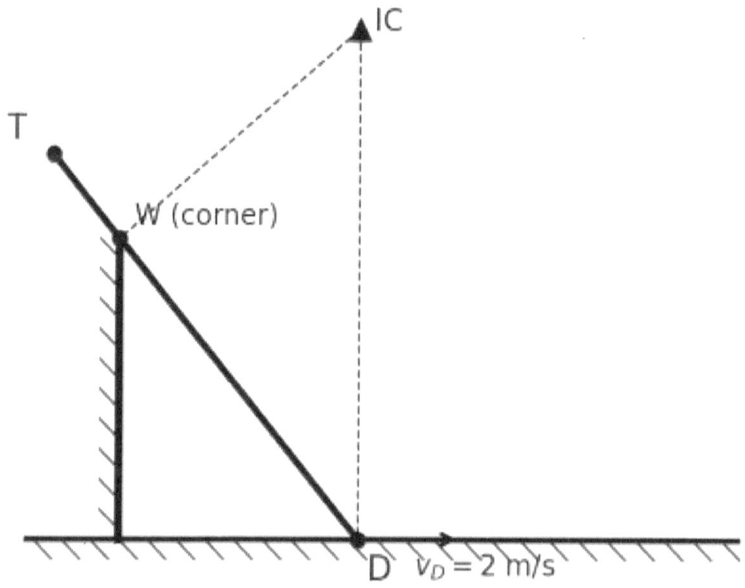

C3.1 (Mechanics Set 3) — Slider-crank mechanism

Crank OA (80 mm) rotates at 3000 rpm, driving connecting rod AB (280 mm) and piston B, shown at the analysed crank angle of 50°.

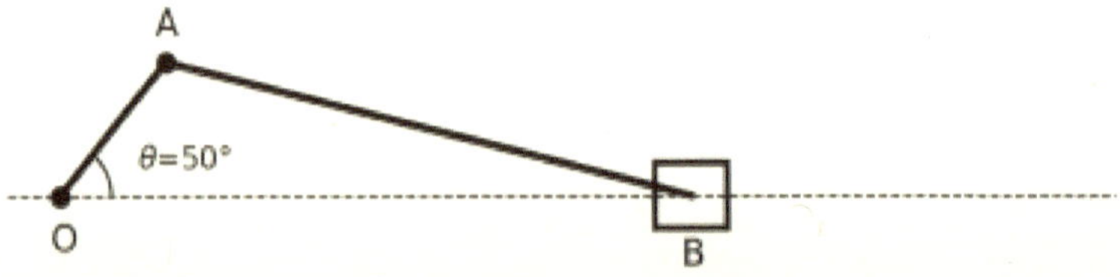

Fig. C3.1 — slider-crank mechanism: OA=80 mm, AB=280 mm, 3000 rpm

C4.2 (Mechanics Set 4) — Particle inside a smooth conical vessel

The particle is projected horizontally at A (height 0.8 m above the vertex) and just reaches the rim at B (height 2.5 m); angular momentum about the vertical axis is conserved throughout because the normal reaction from the cone surface exerts no torque about that axis.

Fig. C4.2 — particle projected inside a smooth conical vessel

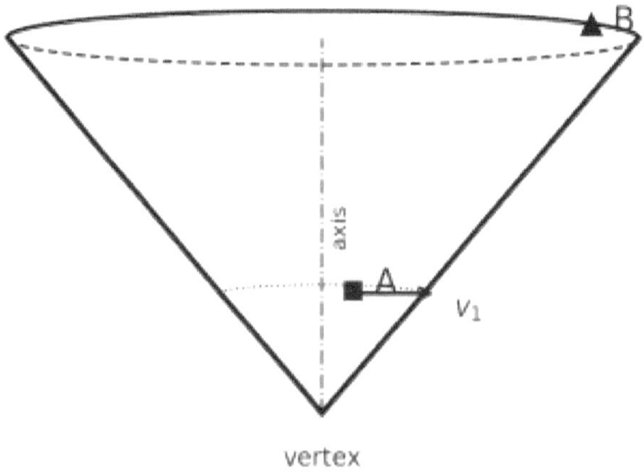

C4.1 (Mechanics Set 4) — Particle on a table linked through a hole

Particle A slides on the frictionless table while particle B hangs below and moves vertically, coupled through the taut string passing through the smooth hole; angular momentum of A about the hole is conserved since the string tension is purely radial.

Fig. C4.1 — particle A on a table, linked through a hole to hanging mass B

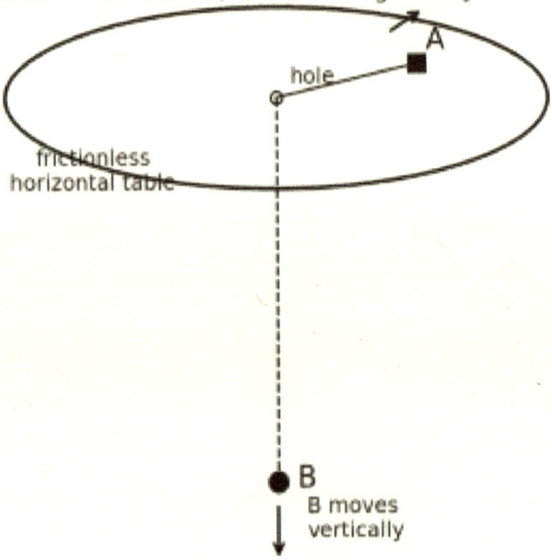

www.ingramcontent.com/pod-product-compliance
Lightning Source LLC
LaVergne TN
LVHW100201061126
851857LV00010B/346
* 9 7 9 8 2 3 5 3 7 1 2 5 5 *